Baby MOOSE

Ellen Lawrence

Only male moose grow antlers.

Photography © Anders Lindgren, Public Domain, Shutterstock 2021 Alex Kednert; astudio; astudio; BrAt82; DGIM studio; DistanceO; DSBfoto; Flas100; grop; hchjjl; Inka1; Intellegent Design; Irmun; Juhku; Kaewta; Kamieshkova; kovalto1; Macrovector; Milanares; MM_photos; Nik Merkulov; Olex Runda; optimarc; Oxlock; Pyty; sociologas; tristan tan; vasabii; VolodymyrSanych; xpixel; Yaska; ZiaMary, Superstock

Published by Sequoia Kids Media,
an imprint of Sequoia Publishing & Media, LLC

Sequoia Publishing & Media, LLC.,
a division of Phoenix International Publications, Inc.

8501 West Higgins Road
Chicago, Illinois 60631

© 2023 Sequoia Publishing & Media, LLC
First published © Ruby Tuesday Books Limited 2021

Customer Service: cs@SequoiaKidsBooks.com

Sequoia Kids Media and associated logo are trademarks and/or registered trademarks of Sequoia Publishing & Media, LLC.

Active Minds is a registered trademark of Phoenix International Publications, Inc. and is used with permission.

All rights reserved. This publication may not be reproduced in whole or in part by any means without permission from the copyright owners. Permission is never granted for commercial purposes.

www.SequoiaKidsMedia.com

ISBN 978-1-64996-218-8

Baby MOOSE

TABLE OF CONTENTS

A New Baby in the Forest 4
Fact File 22
Glossary 24
Index 24
Read More 24
Visit Us 24

Words shown in **bold** in the text are explained in the glossary.

A NEW BABY IN THE FOREST

FOREST

Among the trees in a forest, there lives a baby animal.

The baby has enormous ears and a coat of soft brown hair.

Who does this little baby belong to?

MOTHER MOOSE

NEWBORN CALF

The baby, or calf, belongs to a mother moose.

The little calf was born
on a sunny spring morning.

His mother gently licked him clean.

By the afternoon,
the calf was able
to stand.

At first, the calf's long, thin legs were a little wobbly.

By the end of his first day, however, the baby moose could walk.

A one-day-old moose calf

A one-week-old moose calf

Within a few days, he could run—fast!

The calf stays close to his mother at all times.

The mother moose keeps watch for bears and wolves that might hurt her baby.

MOOSE CALF

MOTHER MOOSE

If a **predator** comes close, she kicks it with her large hooves.

The calf drinks milk from his mother. Sometimes, he drinks standing up.

A CALF DRINKING MILK

Sometimes, the calf and his mother sit down when it's time to feed.

NATIONAL PARK ADVENTURE

MOTHER MOOSE

The mother moose eats plants.

She munches on leaves, new shoots, twigs, and tree bark.

When he is about two weeks old, the calf tries this grown-up food.

MOOSE CALF

MOOSE CALF

The mother moose also likes to eat water plants that grow in lakes and streams.

When mom wades into a lake,
the calf goes, too.

As the summer passes by,
the calf grows bigger and stronger.

MOTHER MOOSE

When he is about one year old, the calf is ready to live alone and take care of himself.

Soon, bony **antlers** start to grow from his head.

ANTLERS

An 18-month-old moose

ANTLERS

The antlers grow and grow.

When he is six years old,
the moose is all grown up.

Now he has enormous antlers.

He weighs more than
1,000 pounds (454 kg).

He has become a tall,
powerful adult bull moose.

ANTLERS

An adult bull moose

FACT FILE
All About Moose

Moose are the largest members of the deer family. They are also known as elk.

Only male moose grow antlers.

Moose are fast runners and strong swimmers.

Each year, a moose's antlers fall off in winter. Then a new pair grows in spring.

A moose's antlers can measure 6 feet (1.8 m) from tip to tip.

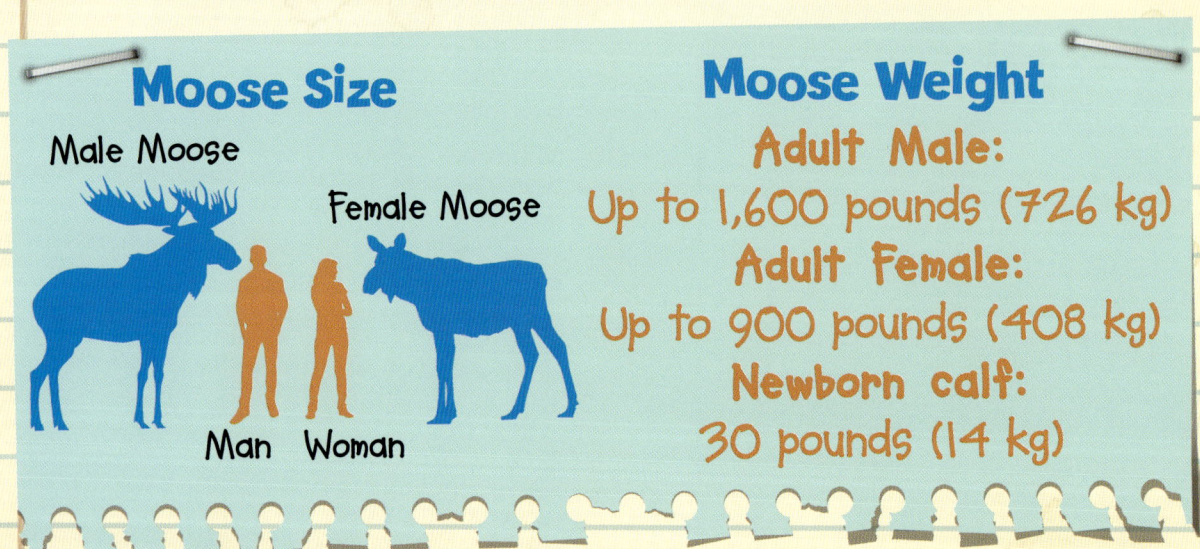

Moose Size
- Male Moose
- Female Moose
- Man
- Woman

Moose Weight
Adult Male: Up to 1,600 pounds (726 kg)
Adult Female: Up to 900 pounds (408 kg)
Newborn calf: 30 pounds (14 kg)

Where Do Moose Live?

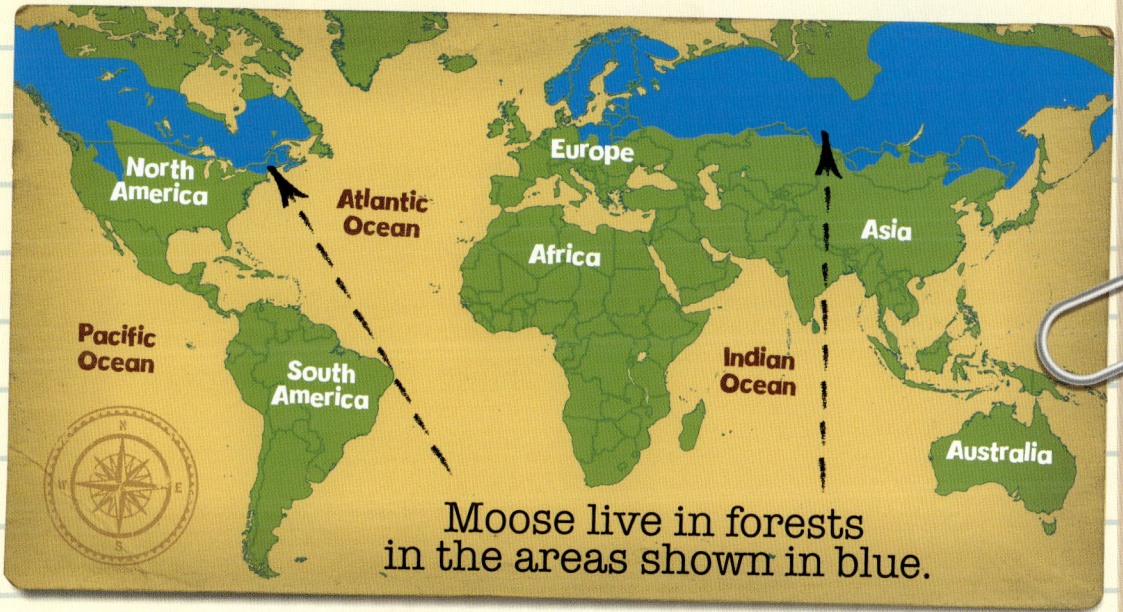

Moose live in forests in the areas shown in blue.

Moose Moms and Dads

Adult moose live alone.
Males and females meet up when it's time to **mate**.
A female moose is ready to have a baby when she's about two years old.

A MOTHER MOOSE AND CALF

Sometimes a female moose gives birth to twin calves.
Father moose do not help take care of their calves.

Glossary

antlers (ANT-lurz)
Large, branchlike body parts made from bone that grow from a deer's head.

mate (MATE)
To get together to have babies.

predator (PRED-uh-tur)
An animal that hunts and eats other animals.

Index

antlers 18-19, 20, 22
food 12-13, 14-15, 16
forests 4, 23
male (bull) moose 20-21, 22-23
mother moose 6-7, 10-11, 12-13, 14, 16-17, 23
predators 11
running 9, 22

Read More

Moose
(My First Animal Library)
Cari Meister
Minneapolis, MN: Jump!
2019

Growing and Changing: Let's Investigate Life Cycles
(Get Started with STEM)
Ruth Owen
Minneapolis, MN: Ruby Tuesday Books
2017

Visit Us

www.SequoiaKidsMedia.com

Downloadable content and more!